PLC Logics and HMI Screens

for

Machine Sequencers Automation

A pratical approach to twin and parallel sequencers
using IEC 61131 - 3 Ladder Logic

AUTOMATION RECIPES - Volume 3

Rosario Cirrito

Copyright

All rights reserved. No part of this publication may be reproduced, stored in a retrieval system, or transmitted, in any form or by any means, electronic, mechanical, photocopying, recording or otherwise, without the prior permission of the author.

Every effort has been made to make this book as accurate as possible, however, there may be mistakes, both typo and in content. This content should be used as a guide, the result of a thirty-year experience of the author as a PLC - HMI - SCADA developer.

Suggestions, comments and requests for explanations or details are welcome, please send them to:

author.rosario.cirrito@gmail.com

Note: This book contains many images. Since eReaders don't always display images well, I would like to provide you with the PDF file that contains this book and the images will be easier for you to see. To receive PDF version of this book, you just need to email me prove purchase of kindle version of the book from Amazon.com to author.rosario.cirrito@gmail.com. Upon receive of that, PDF version will be emailed to your mail box.

ISBN 9781980704706
Independently published

Synopsis

This booklet is the third of a series dedicated to automation recipes created with the PLC (Programmable Logic Controller) and HMI (Human Machine Interface) binomial. The series is aimed at an audience of readers with an elementary knowledge of PLC programming, eager to learn advanced solutions, extensively tested on real systems.

In modern computer programming, generally oriented to the development of "object-oriented" software, the developer strives, as much as possible, to resort to so-called "Design Patterns", standard solutions for frequently recurring problems: A design pattern describes a problem, particularly recurring in a given context, and then provide the heart of the solution to this problem. It is therefore possible to successfully reuse this solution, thousands and thousands of times, with the certainty of using an efficient and well-tested solution.

In the present series, which deals exclusively with development on PLC-HMI, the term "design pattern" has been replaced by the term "automation recipe" for an easier understanding by the non IT reader.

In the chapters of this book we will show in detail an automation recipe that can be reused in any PLC-HMI automation project that uses "electric motors". The recipe has also been optimized for operation with Scada supervision systems.

The first book deals with electric motors automation while the second with measurement and monitoring quantities acquired with 4-20 mA current sensors.

This third book deals with the topics of sequencers for starting / stopping machinery such as pumps or compressors that operate in twin or parallel mode on groups or technological plants.

The sequencers process the signals output from a monitoring block of an analogue quantity and send the start command to a certain number of electric motors, which are automated with electrical blocks of type ElectricMotor (see volume 1). They therefore complete the regulation chain of an analogue quantity with digital drives (start / stop).

In detail, the first section of this booklet, dedicated to the application domain, analyzes the two types of sequencer: twin for the operation of two machines, one of which is always on stand-by or parallel to start / stop a certain number of machines, generally of the same size, installed in parallel.

The second section deals with the development of combined software for both PLC and HMI. The logic of the two functional blocks (UDFB), Mot2Seq and Mot6SEq, and the related display screens, for local monitoring and for setting configuration and timing parameters are illustrated.

Finally, the third section shows the application of the concepts developed in a real level control case in a waste water pumping station.

The HMI solutions have been extensively tested on the OCS, Operator Control System, manufactured by Horner Apg. OCS combines a Controller, Operator Interface, Network and I/O into a single product. While the author, has been widely using Siemens, Allen Bradley, GE Fanuc PLCs he has focused the books of this series on the Horner OCSs because Horner provides Cscape, an integrated development environment, extremely easy to use and above all completely free.

All the logics, published in the book, have been developed using the IEC61131-3 compliant Ladder language; therefore it is extremely easy to migrate them on almost all the PLCs of other manufacturers.

The same applies to HMI screens whose graphic controls are very similar on the different equipment offered on the market.

The reader who already has experience with other manufacturers' equipment can therefore continue to use what he knows best.

Contents

Cover
Title
Copyright
Synopsis

1. The application domain
1.1 The twin sequencer
1.2 The parallel sequencer

2. The application PLC - HMI software
2.1 The modular programming and the memory mapping
2.2 The twin sequencer function block
2.3 The twin sequencer human machine interface
2.4 The parallel sequencer function block
2.5 The parallel sequencer human machine interface

3. A real world example
3.1 Water level measurement and control logics
3.2 Water level measurement and control HMI
3.3 Pumps sequencer logic
3.4 Pumps sequencer HMI

4. Summary

1.1 The twin sequencer

In technological systems it is often necessary to guarantee a reserve function for a certain machine.
A typical example is given by the twin pumps of the secondary circuits of the hydronic water systems such as those shown in figure 1.1.1:

The use of a twin unit or two single pumps, of the same characteristics, as those shown in figure 1.1.2, is dictated by the need to be able to start the second pump in case of failure of the first:

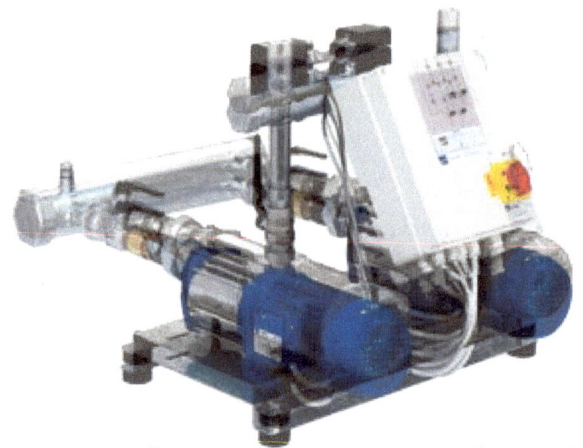

The choice of a twin group rather than two single pumps is justified by a simplification on the plant that is reflected in the installation costs.
In the case of groups with two single pumps, a delivery manifold must be provided and two intersecting valves and one non-return valve on each pump must be provided.
With the twin group, only two valves are needed, one for suction and one for each group. A considerable reduction in overall dimensions is also obtained so as to allow the realization of hydronic circuits with a single delivery manifold from which multiple twin groups branch off, as shown in Figure 1.1.3:

From the functional point of view one of the two pumps is always in reserve to the other; simultaneous operation is generally not required.

A secondary problem, but not for this less important, that this type of sequencer is called to solve, is to guarantee the cyclical alternation of the starts in order to ensure that the two pumps work roughly for the same number of hours per year in order to guarantee uniform wear.

1.2 The parallel sequencer

The parallel sequencer performs the task of starting / stopping a certain number of machines, generally of the same size, in order to maintain a certain physical quantity, for example level or pressure, around a value set by the operator.
The machines are therefore installed in parallel in pre-assembled sets. A typical example is the water booster set (see Fig. 1.2.1) for drinking or fire-fighting water

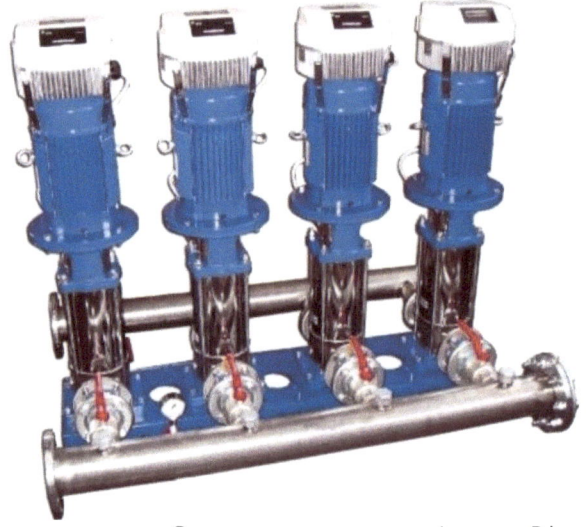

or the group of compressors (see fig.1.2.2) for refrigeration systems.

The parallel sequencer can also be used, in applications with significant installed electrical power, as in the case of screw compression units for air conditioning units or for the machine rooms of refrigeration systems or for compressed air units.

In the air-conditioning systems, the individual cooling / heating units are switched on / off in the primary hydronic circuits in order to maintain the flow temperature between a minimum and maximum prefixed value.

In refrigeration systems, the compressors are switched on / off in order to keep the evaporation pressure and therefore the minimum temperature of the refrigerating cycle as high as possible at the preset value.

In the compressed air units, the screw units keep the compressed air distribution pressure at the plant as constant as possible.

2.1 The modular programming and the memory mapping

The types of languages provided by the IEC 61131-3 standard can be considered as low-medium level programming languages.
This standard was developed to ensure a certain portability of the programs between PLCs from different suppliers.
Its greatest merit consists in being oriented to the modular development of control applications thus allowing the overall logic to be divided into subroutines recalled cyclically by a single main program.
The individual subroutines can in turn recall standard functional blocks provided by the language or even functional blocks UDFB, acronym of User Defined Function Block, expressly developed by the user.
A function block groups an algorithm and a set of private data so it is parameterized with reference to the input and output variables; this allows the developer to create multiple instances of each block belonging to the same automation system:
We will develop this important concept in greater detail in the following paragraphs.
To program subprograms and function blocks, the standard allows the developer to use one of the following five languages:
1) Ladder Diagram (LD)
2) Instruction List (IL)
3) Function Block Diagram (FBD)
4) Structured Text (ST)
5) Sequential Function Chart (SFC)
The choice is dictated by personal preferences or by the specific professional background of the programmer.
The modular decomposition of the application is clearly visible in Figure 2.1.1 which shows the structure of the various components typical of a project realized with the Horner CScape development environment.

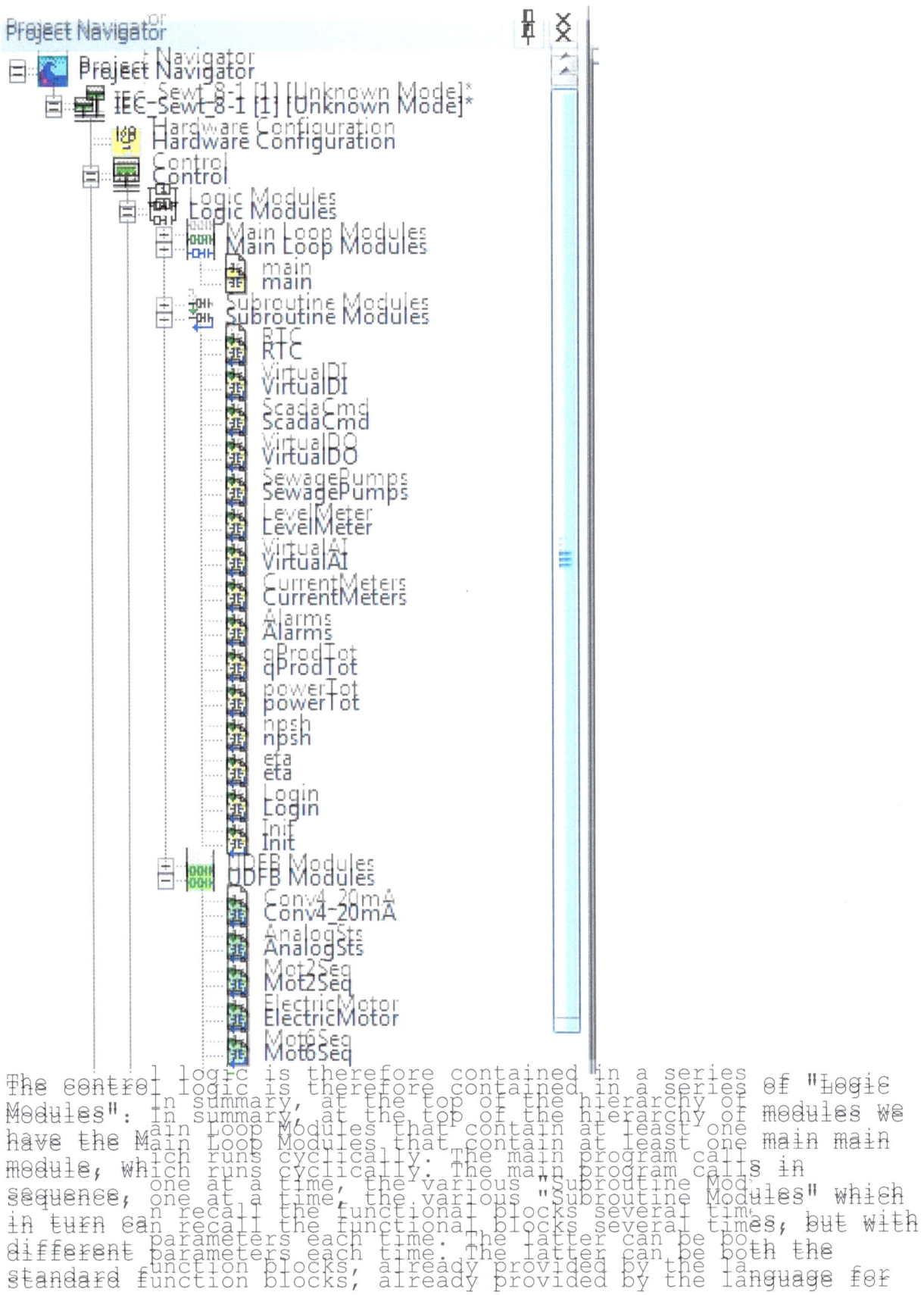

The control logic is therefore contained in a series of "Logic Modules". In summary, at the top of the hierarchy of modules we have the Main Loop Modules that contain at least one main module, which runs cyclically. The main program calls in sequence, one at a time, the various "Subroutine Modules" which in turn can recall the functional blocks several times, but with different parameters each time. The latter can be both the standard function blocks, already provided by the language for the general purpose logical instructions, or the specific user-defined blocks, the "UDFB Modules".

The PLC control logic as well as the HMI display logic is however developed on a PC, almost exclusively in the Windows environment. The relevant source file is saved, periodically and after complete modifications, on the PC hard disk.

Whatever the hardware being used, PC or PLC, a set of RAM working memories is always needed both to store the program instructions and to save, at each scan cycle, the data of the dynamic variables. Today's PC generally has a RAM of 4-8 GBytes, while the PLC requires far more modest memories, from 256 kB to 1MB, to store control logic of even particularly complex systems, as well as a few thousand internal variables.

High-level PC languages use Short, Byte, Integer, Long, Float, Double primitive variables that occupy from 8 to 64 bits of memory; the data type most frequently used by the PLC is instead the Word, composed of 16 bits, also called register (% R). A single Word or Register, having 16 bits in total, thanks to the binary numbering system, can represent Integers with a sign between -32768 and + 32767 or without sign in the range 0 and 65365.

When it is necessary to represent integers of higher value or real number, a 32-bit representation is used, obtained by merging two adjacent 16-bit registers.

A 16-bit register can also be used to aggregate the binary state of logical bits, each of which occupies one bit, in groups of 16. This packed solution is particularly compact and efficient especially when these variables are transmitted to Scada systems or transferred on the network from one PLC to another.

The individual bits of the Boolean variables then become individually accessible at address% Rx.y with x, index of the register, and y index of the bit, between 1 and 16: so %R1.5 will indicate the bit 5 of the register 1. Boolean binary values can in any case be stored also as retentive variables of type% M or non-retentive of type% T.

In addition to storing integers, real numbers and packed bits, the% R registers are also used to store enumerations of machine and sensor states that can be associated with predefined text strings in the HMI interface screens. We will show such use when we take care of displaying dynamic texts in the HMI panel.

A memory representation of a physical quantity, acquired in real time, is, for example, a pressure that at a certain moment takes on a value equal to 8.95 bar. In this case we can represent it as a 32 bit real value, using two consecutive registers, for example %R201 and %R202; or as an integer value, equal to 895, with occupation of a single 16-bit register, for example% R200.

This second mode allows us to store the actual data in half the memory space, which is especially important when the data must be sent to a supervision system or another controller along a serial line that is not too fast; but this approach has the drawback that the correct representation of the displayed format must always be managed, by PLC logic and HMI configuration, keeping in mind how many decimal digits are to be taken into account.

An example of a register containing an enumeration of dynamic texts is constituted by the status register of a pump whose value can vary, in real time, within a certain set of logic precoded states stored in tabular form within the HMI device, as shown in figure 2.1.2.

Value	Text
0	???
1	ON
3	ON_SEL
4	OFF
5	REM_0
6	LOC_0
18	ALARM
32	INIBIT
64	INTERD
130	FDBACK

The integers shown in the Value column correspond to the logical states shown in the Text column. The latter can therefore be displayed in a text field inside the graphic pages of the operator panel associated with the PLC.

2.2 The twin sequencer function block

Problem

In the water systems the presence of two twin pumps is very widespread, one of which has only the function of reserve for the other. A classic example is the emptying of waste and / or meteoric water with two pumps. In order to equal wear, the pumps are started, one at a time, alternately, while in exceptional cases, corresponding in general to exceeding alarm thresholds, both are started or stopped.

Solution

The Mot2Seq twin sequencer strategy dialogues with both ElectricMotor modules of the two pumps, described in detail in the first book of this series, sending them the start command via the boolean variable Go and receiving the information of On, Ready and FbNok.

The control logic of the design pattern is contained in the UDFB module named Mot2Seq. The functional block will be called, within the main program, for each group of twin or twin pumps to be driven.

Map of local variables

The table of input, output and internal variables of the block is shown in figure 2.2.1:

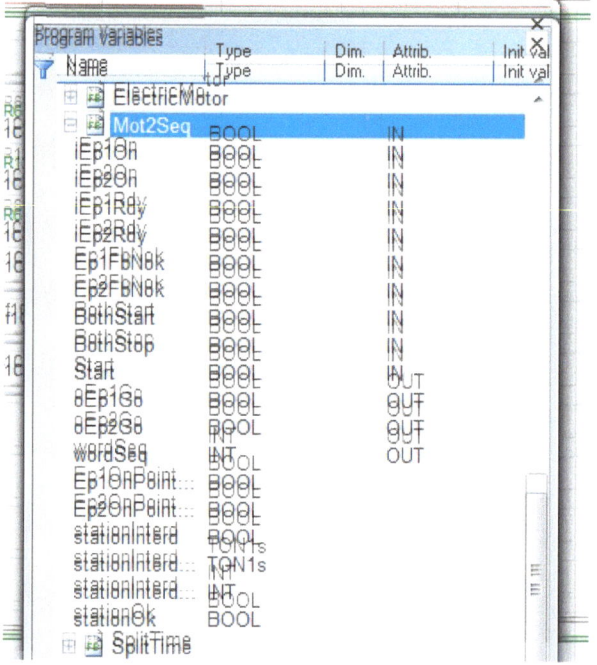

For each of the two controlled pumps the Boolean variables of On, ready to start (Rdy), and no feedback state (FbNok) are acquired.
Then there are three other Boolean input variables that respectively indicate: BothStart the need to start both pumps, BothStop to stop both, and finally Start indicating the need to implement the twin sequence strategy.
The output variables are three: the two Boolean variables of Go for the start of each pump and the whole 16-bit wordSeq variable that indicates the status of the sequencing station, used to monitor the correct functioning of the sequencer.

The Mot2Seq module

An example of EpCondSeq instance, directly calle by the main module for the alternation of two condensation pumps of a refrigeration system, is shown in Figure 2.2.2:

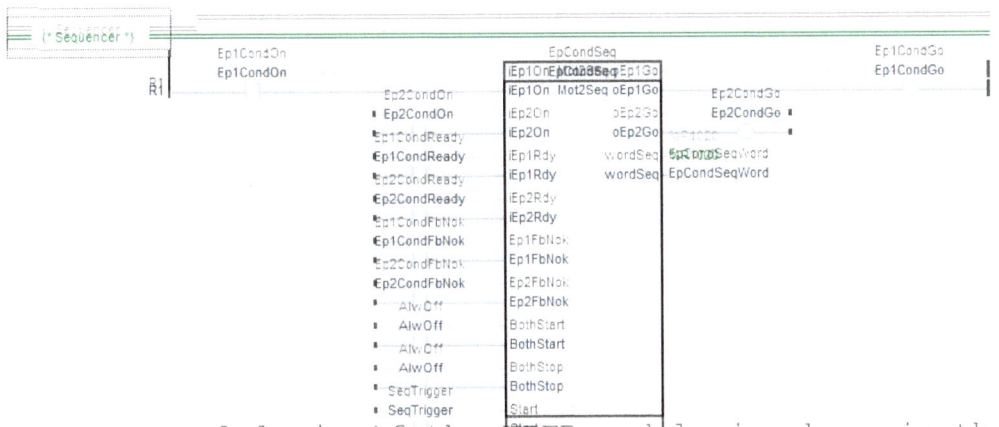

The internal logic of the UDFB module is shown in the following rungs:
Rungs R1=R2, as shown in figure 2.2.3, reset the stationInterd interdiction flag by means of a standard delay timer. This timer will be activated, via the aforementioned flag, every time the strategy takes any action on the controlled pumps. Setting this timer to 10 seconds means that these actions can not be taken too frequently. If desired, the block can be easily modified by adding the value of the interdiction timer set to the input parameters:

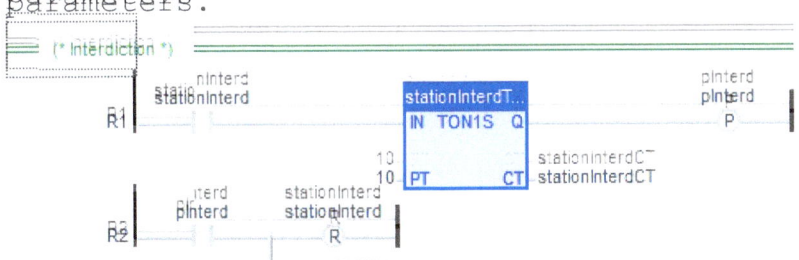

The rungs R3-R4 command, in the presence of the external signal of BothStart, the Go of both pumps, also setting to 2 the value of the wordSeq and returning the logic flow to the main program, as shown in figure 2.2.4.

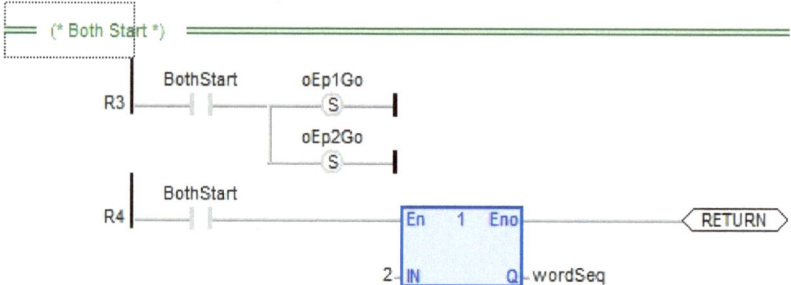

The R5-R6 rungs control, in the presence of the external signal of BothStop or that of lack of Start, the stop of both pumps (resetting the relative Go), also setting to 3 the value of the wordSeq and returning the logic flow to the main program, as shown in figure 2.2.5.

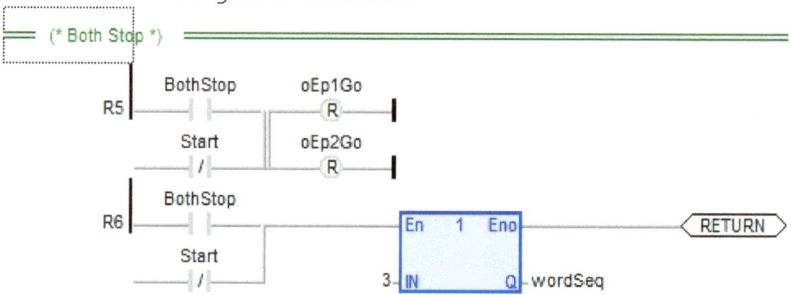

The rung R7, when there is an interdiction flag, set the wordSeq to 1, returning the logic flow to the main program, as shown in figure 2.2.6.

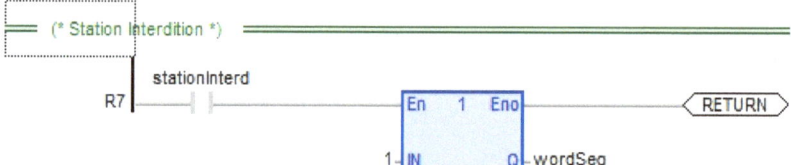

The R8-R9 rungs verify that a pump is started and not in a failed state while the other is stopped. In this case the block recognizes that everything is OK and therefore sets the value of the wordSeq to 0 and returns the control to the main program, as shown in figure 2.2.7.

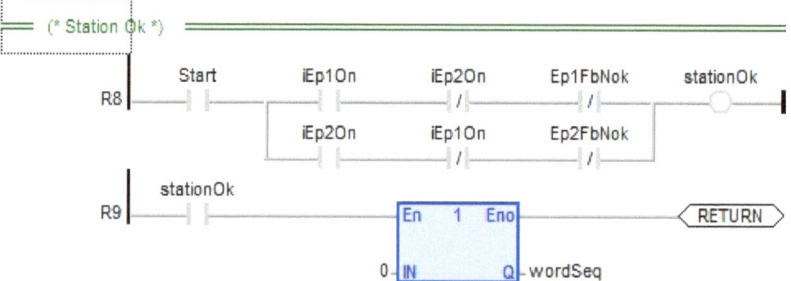

If rung R9 is not satisfied, the logic flow continues with line R10 which, if pointers are found both set to 0, set the first start pointer, as shown in figure 2.2.8.

Rung R11 checks whether there are the conditions to start the first pump, i.e. if the first pump is pointed and if it is ready to start and not in FeedBack state not Ok. In this case, the Go is given to the first pump, the consent to the second is denied, the pointers for the next On are updated and the interdiction flag is set.

Rung R12 sets the value of wordSeq to 4 if the previous line has energized the stationInterd flag and returns the flow to the calling program.

Otherwise the line R13 proceeds to update the pointers of the pumps to be started, as shown in figure 2.2.9.

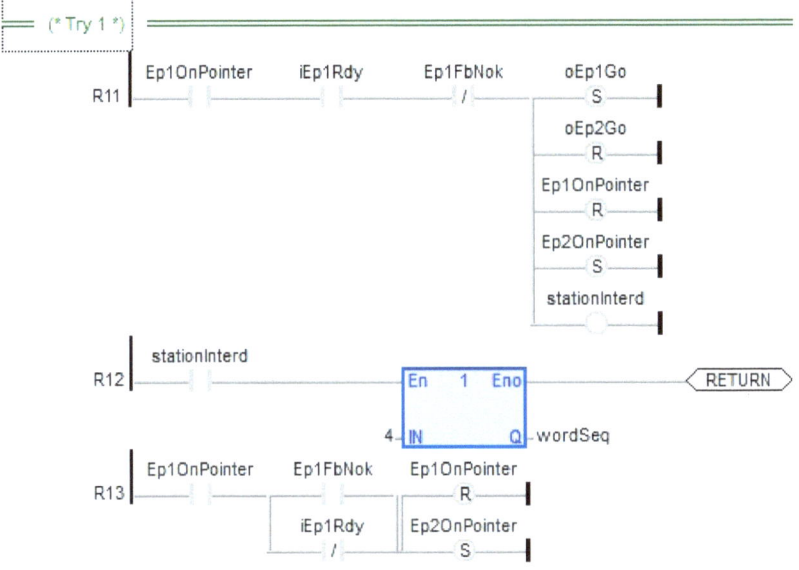

The R14-R16 rungs operate like the R10 and R12 lines but on the second pump, as shown in figure 2.2.10.

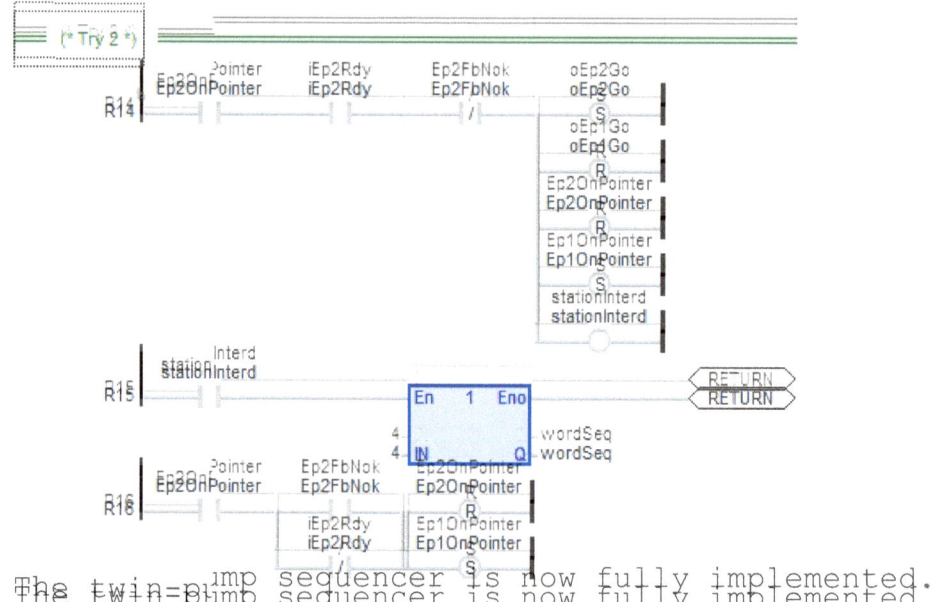

The twin-pump sequencer is now fully implemented.

2.3 The twin sequencer human machine interface

A real world example of a screen displaying a station with a twin pump sequencer for an evaporative cooling tower of a refrigeration system is shown in Figure 2:3:1:

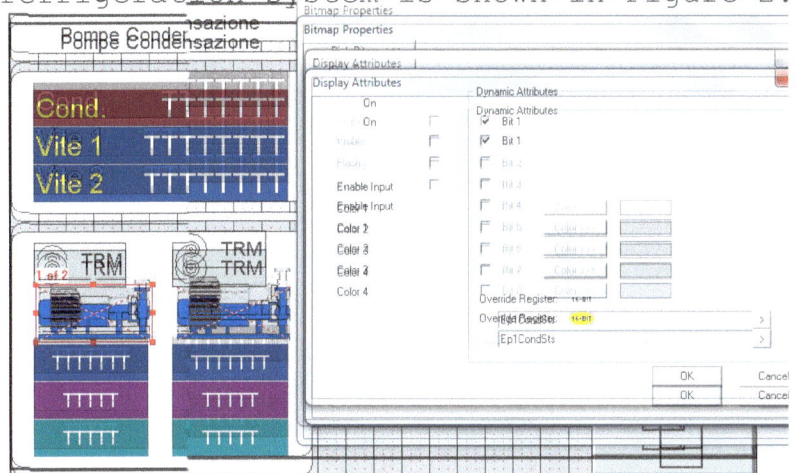

The pump symbol is only visible if it is On; only in this condition the bit 1 of the state variable associated with the pump is active (see block UDFB ElectricMotor in the booklet 1 of the series):

2.4 The parallel sequencer function block

Problem
The need to start / stop a certain number of machines, operating in parallel, in order to regulate a process variable is perhaps the most frequent request for various types of systems.
In water booster systems, a typical pumping set provides 2 to 6 electric pumps that are switched on / off in order to regulate the water pressure on the delivery pipe. In the waste water lifting stations the same principle is used, with a number of pumps installed varying from 2 to 4 pumps, to control the level in the receiving tank.

Solution
The sequencer we will develop can drive up to 6 parallel operating machines. The number of machines actually controlled can be set as input parameter NrPmp. The sequencer will dialogue with the ElectricMotor modules (see notebook 1 of the series), of each machine, sending them the start command via the boolean variable Go and receive the information of On, Ready and FbNok. The control logic of the design pattern is contained in the UDFB module named Mot6Seq. The function block will be called, in the main program, for each group of pumps to be driven.

Map of local variables
The block input variables table is shown in figure 2.4.1:

Mot6Seq		
iEpOn1	BOOL	IN
iEpOn2	BOOL	IN
iEpOn3	BOOL	IN
iEpOn4	BOOL	IN
iEpOn5	BOOL	IN
iEpOn6	BOOL	IN
iEpRdy1	BOOL	IN
iEpRdy2	BOOL	IN
iEpRdy3	BOOL	IN
iEpRdy4	BOOL	IN
iEpRdy5	BOOL	IN
iEpRdy6	BOOL	IN
iEpFbNok1	BOOL	IN
iEpFbNok2	BOOL	IN
iEpFbNok3	BOOL	IN
iEpFbNok4	BOOL	IN
iEpFbNok5	BOOL	IN
iEpFbNok6	BOOL	IN
iHiAl	BOOL	IN
iLoAl	BOOL	IN
iHiOp	BOOL	IN
iLoOp	BOOL	IN
NrPmp	INT	IN
Start	BOOL	IN
setInterdition	INT	IN

We have three Boolean variables, for each pump, related to the ON flag, ready to start (Rdy = Ready) and non-status (Feedback not OK = FbNok). Subsequently we had the four flags linked to the high alarm thresholds (iHiAl), low alarm (iLoAl), high operative (iHiOp) and low operative (iLoOp), set at the output by any instance of the AnalogSts function block.

Finally we have the integer variable NrPmp to specify the number of pumps from 2 to 6 to start / stop in sequence, a boolean variable Start to condition the global start / stop of all pumps as well as the whole setInterdition variable used to set the timing of interdiction.

The table of output variables is shown in figure 2.4.2. We have the six output variables corresponding to the GO of each pump, an integer wordSeq variable used for debugging purposes and a boolean variable oSeqInterd that signals the pumping station interdiction condition.

The table of internal variables is shown in figure 2.4.3. We have six pointers that identify the next pump to start EpOnPointer and the corresponding six EpOffPointer pointers for the next pump to be switched off, four variables to manage the ban and a stationOK variable to define the absence of the need to take start / stop actions.

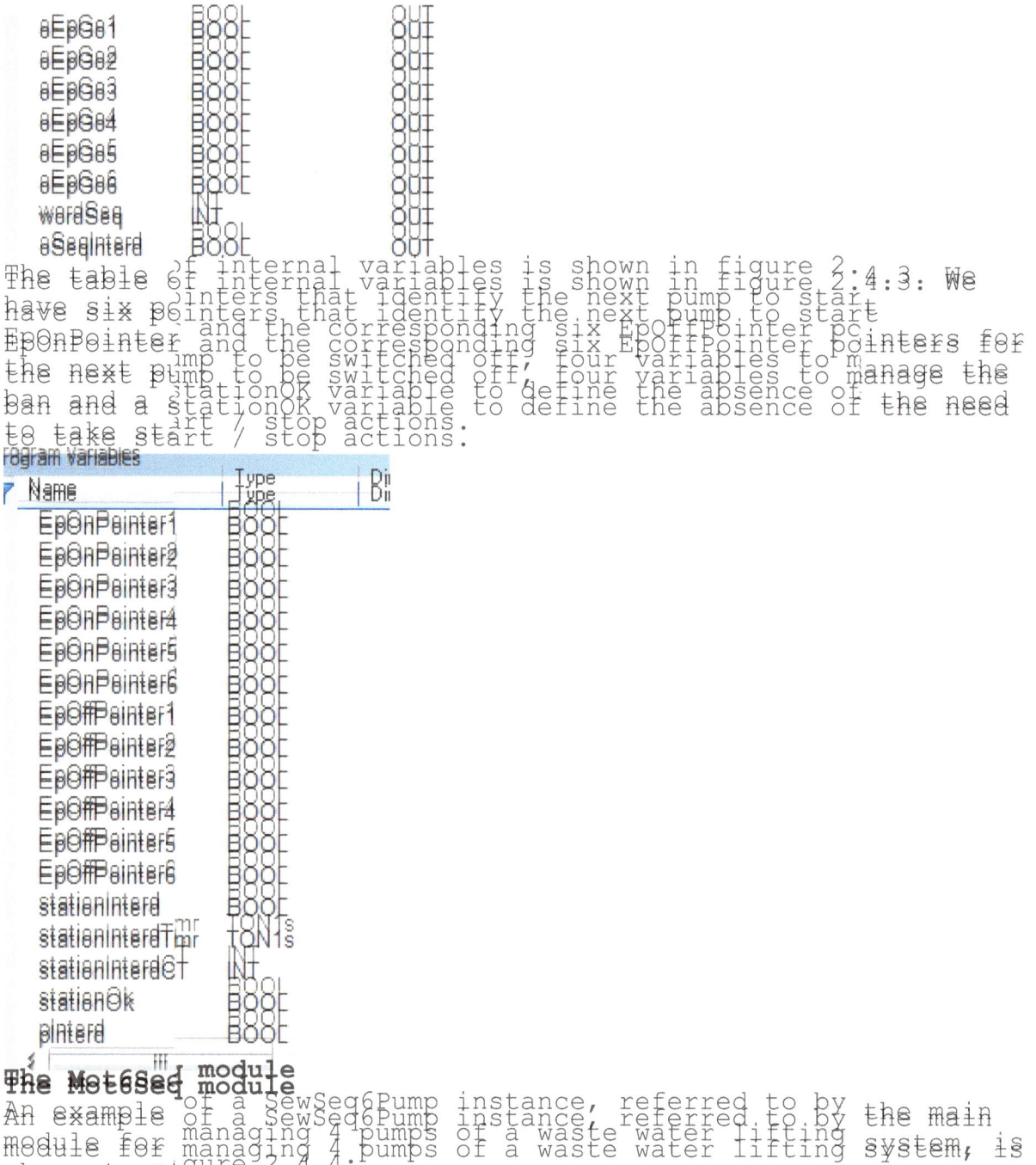

The Mot6Seq module

An example of a SewSeq6Pump instance, referred to by the main module for managing 4 pumps of a waste water lifting system, is shown in Figure 2.4.4:

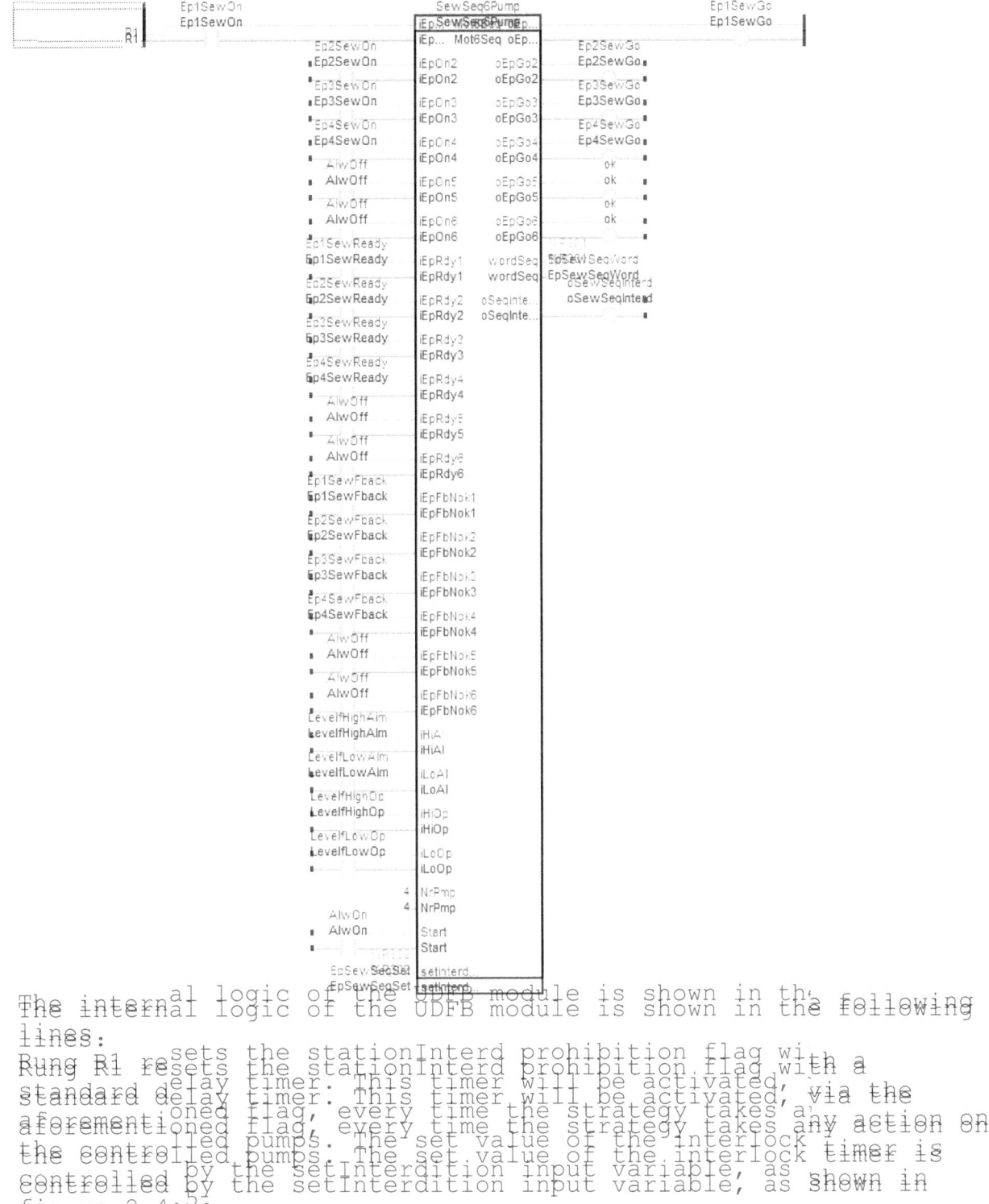

The internal logic of the UDFB module is shown in the following lines:

Rung R1 resets the stationInterd prohibition flag with a standard delay timer. This timer will be activated, via the aforementioned flag, every time the strategy takes any action on the controlled pumps. The set value of the interlock timer is controlled by the setInterdition input variable, as shown in figure 2.4.5:

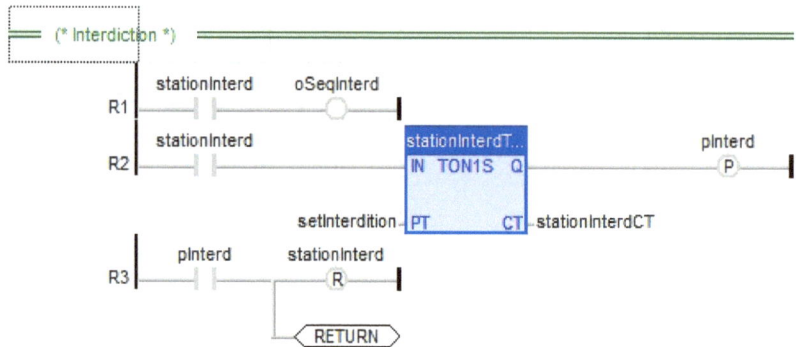

The R4-R5 rungs control, in the presence of the AllStart external signal, the Go of all the machines, also setting the value of the wordSeq to 2 and returning the logic flow to the main program, as shown in figure 2.4.6.

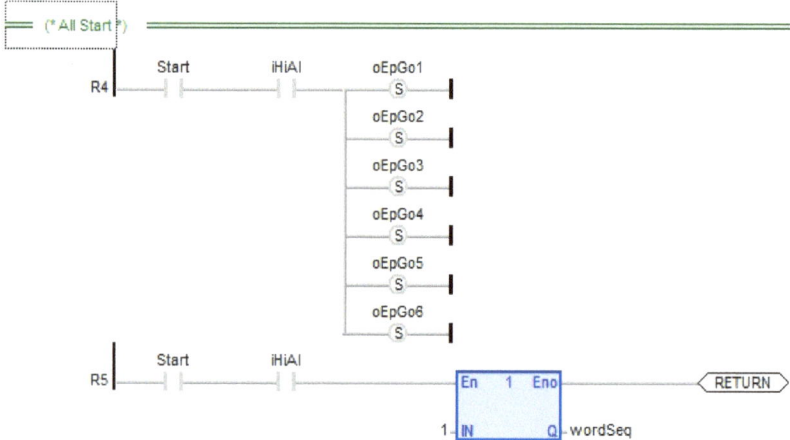

The rungs R6-R8 command, in the presence of the external signal of iLoAl or that of lack of Start, the stop of all pumps (denying the relative Go), setting to 1 the value of the wordSeq and returning the logic flow to the main program, as shown in figure 2.4.7. In the presence of iLoAl, line R7 sets the wordSeq to 2, returning the logic flow to the main program.
The R8 line in the absence of Start sets the wordSeq to 3, returning the logic flow to the main program.

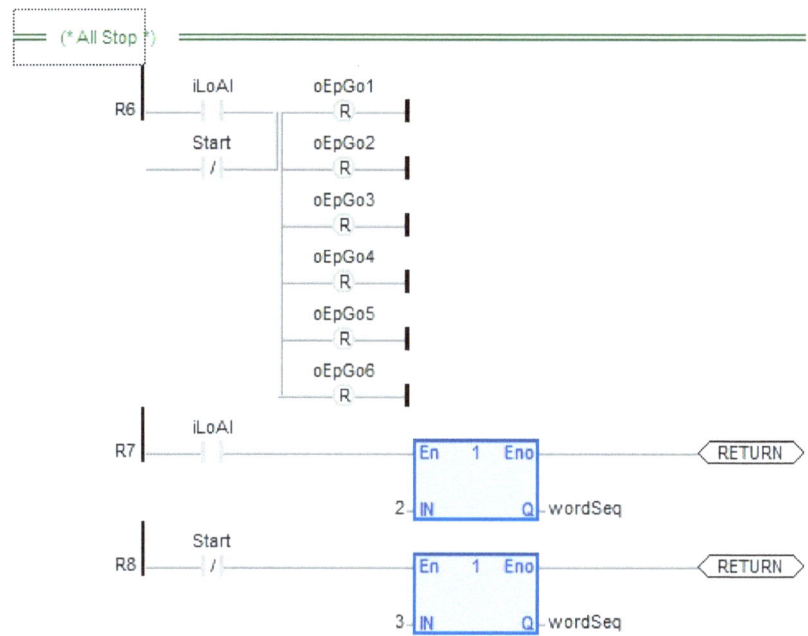

Rung 9 verifies the presence of interdiction by returning the logic flow to the main program, as shown in figure 2.4.8.

The R10-R11 rungs verify that it is not necessary to start / stop any new machinery. In this case, the value of the word Seq is set to 4 and control is returned to the main program, as shown in figure 2.4.9.

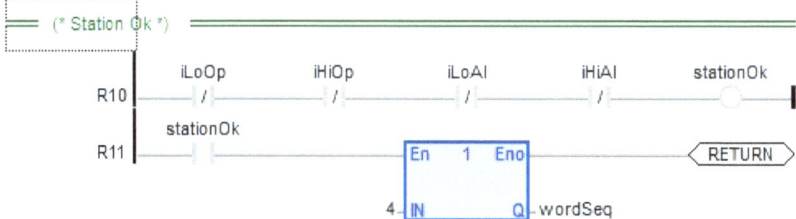

If rung R11 is not satisfied, the logic flow continues with lines R12-13 which, if the pointers are all set to 0, set the first to start / stop, as shown in figure 2.4.10.

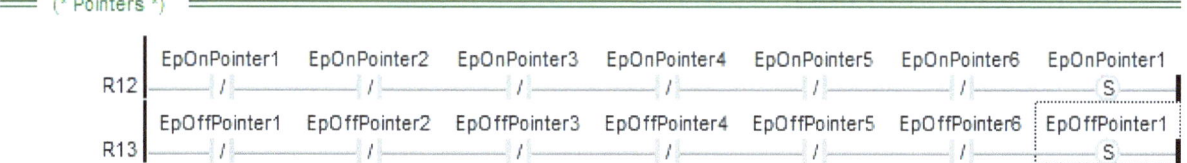

The R14 rung commands a jump to the More label in the event that for the persistence of iHiOp it is necessary to start an additional machine. Line 15 does the same thing in the case of a reduction, as shown in figure 2.4.11.

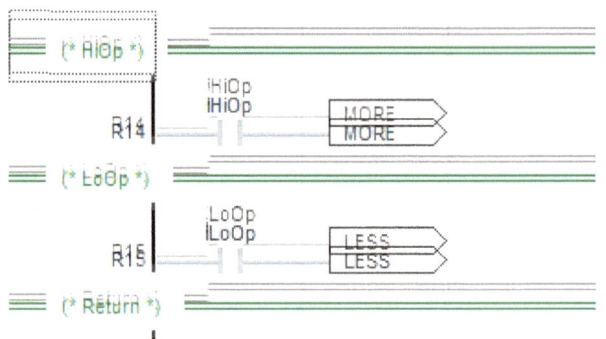

Line R17 checks if there are the conditions to start the first machine, i.e. if it is off, if it is pointed and if it is ready to go and not in FeedBack not OK status. In this case the Go is given to the first machine and the pointers for the next On are updated and finally the interdiction flag is set.

Line R18 sets the value of wordSeq to 5 if the previous line has energized the stationInterd flag and returns the flow to the calling program.

Otherwise the line R19 proceeds to update the pointers of the machines to be started, as shown in figure 2:4:12:

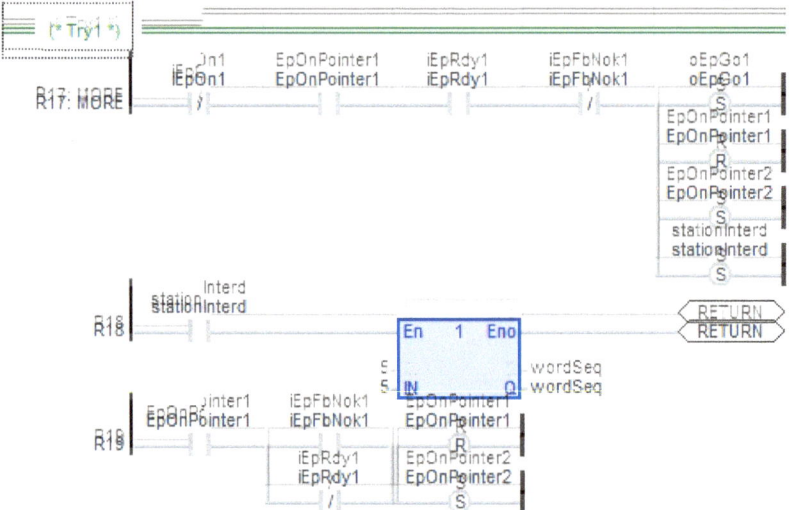

The R20÷23 rungs operate like the previous rungs but on the second machine rather than on the first one. If the total number of machines to be managed is only 2 the first one is set, the pointer flag of the next machine to be started is updated, as shown in figure 2:4:13:

Rungs R24-27 operate like the previous lines on the third machine, as shown in figure 2.4.14:

Rungs R28-31 operate like the previous rungs on the fourth machine, as shown in figure 2.4.15:

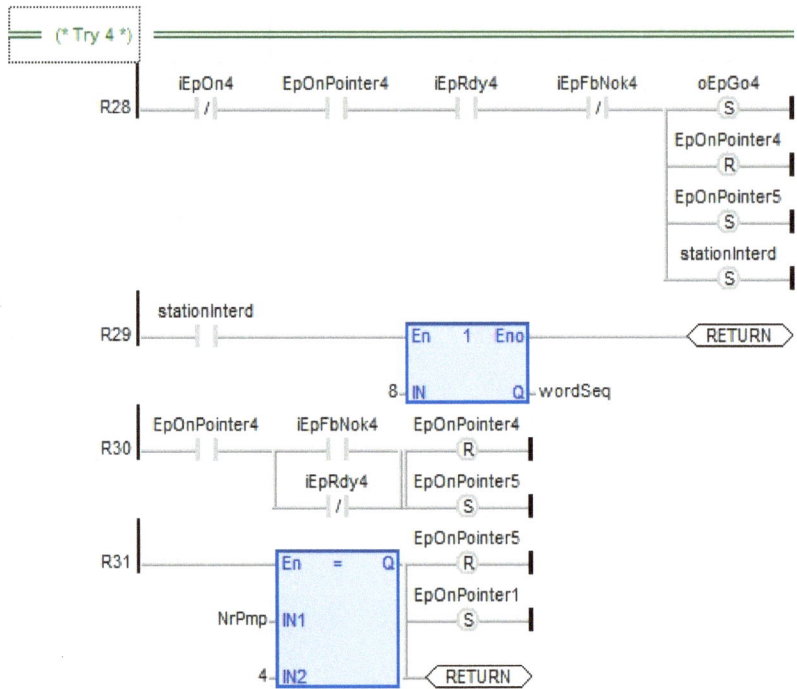

The R32-35 rungs operate like the previous lines on the fifth machine, as shown in figure 2.4.16.

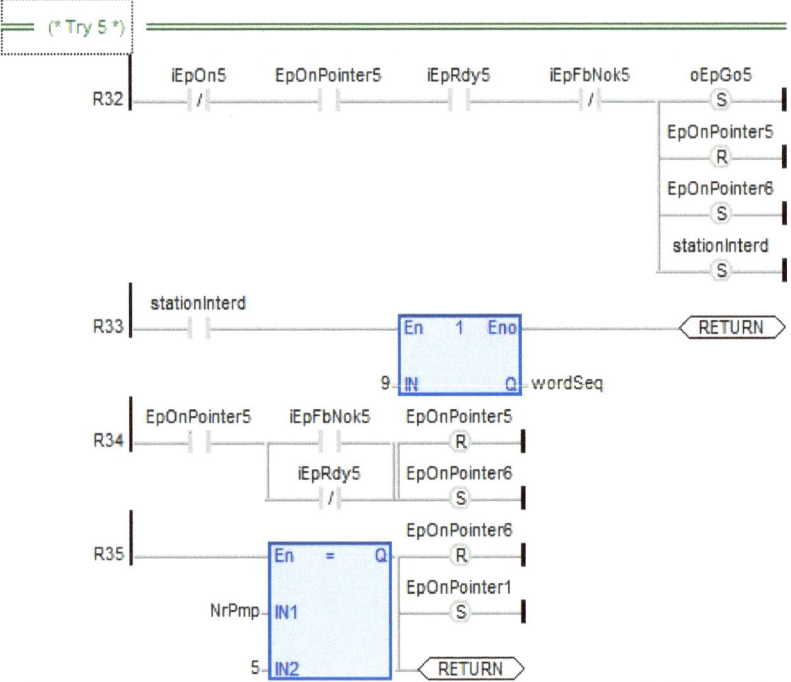

Finally, the R36-40 rungs operate like the previous lines on the sixth machine, as shown in figure 2.4.17.

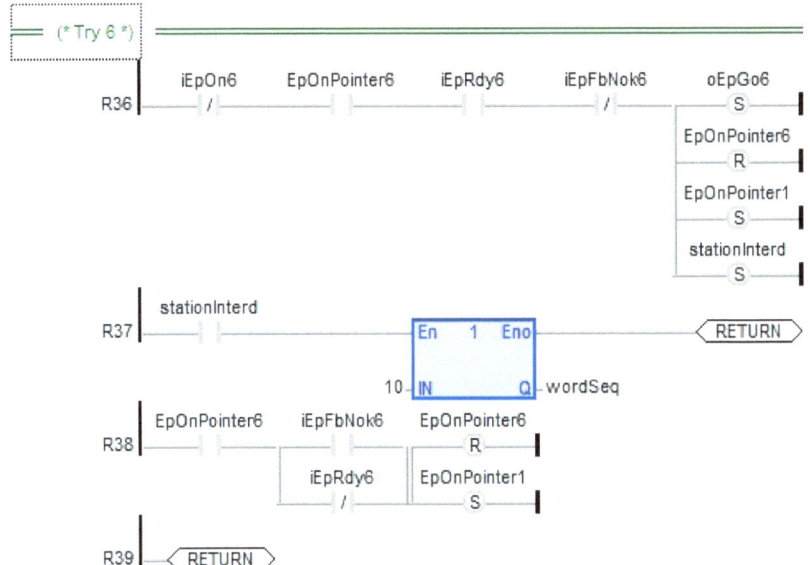

From the R40 rung the reduction process is developed. Rung R40 checks whether there are conditions to stop the first machine, ie if it is started and if it is switched off. In this case, the Go is removed from the first machine and the pointers for the next Off are updated and finally the interdiction flag is set. Line R41 sets the value of wordSeq to 11 if the previous line has energized the stationInterd flag and returns the flow to the calling program.
Otherwise, row R42 updates the pointers of the machines to be started, as shown in figure 2.4.18.

Rungs R43-45 manage the stopping of the second machine, as shown in figure 2.4.19.

Rungs R46=48 manage the shutdown of the third machine, as shown in figure 2.4.20:

Rungs R49=51 manage the stopping of the fourth machine, as shown in figure 2.4.21:

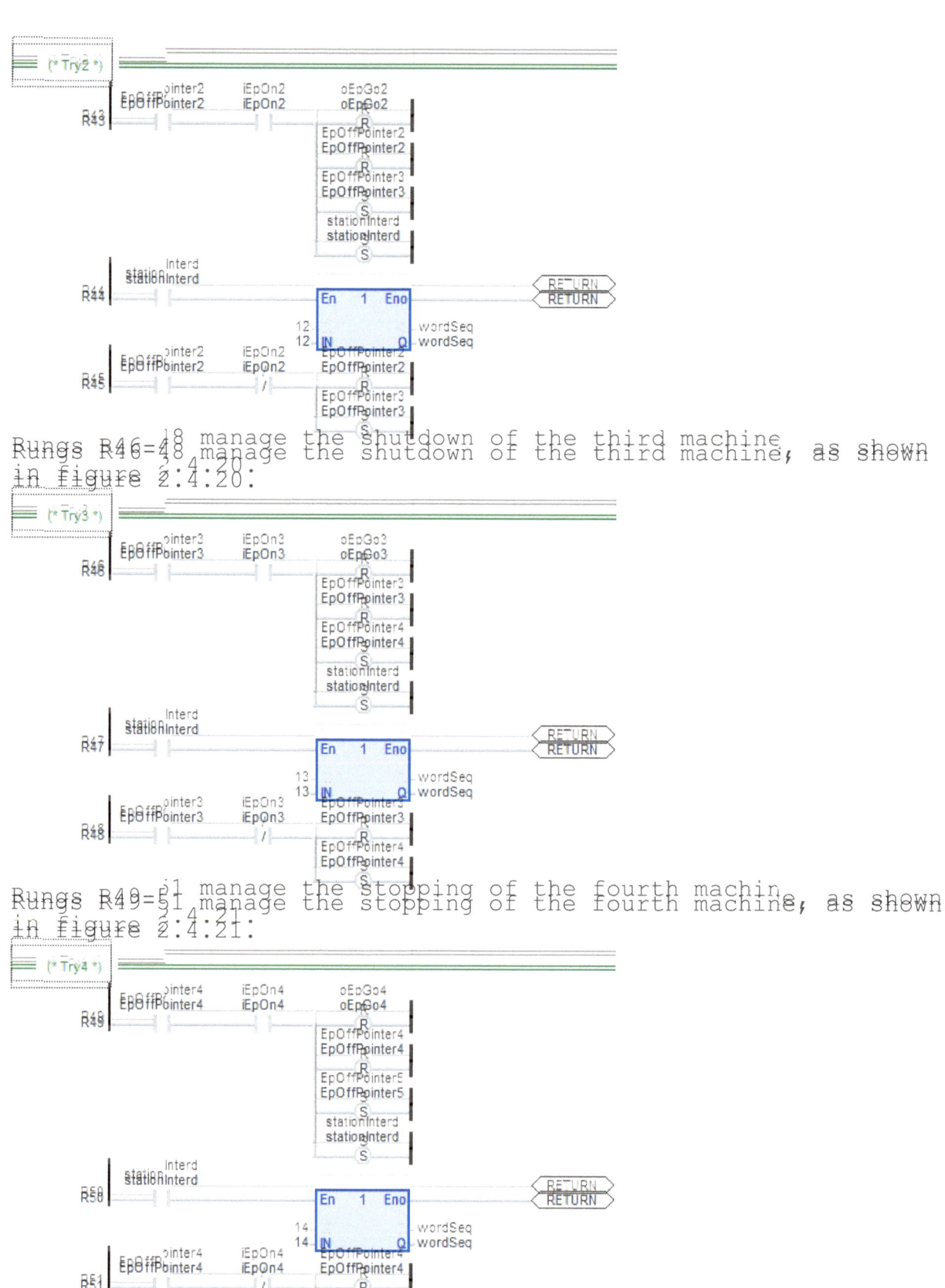

Rungs R52=54 manage the stopping of the fifth machine, as shown in figure 2.4.22:

The R55=57 rungs manage the stopping of the sixth machine, as shown in figure 2.4.23:

The Mot6Seq implementation is completed here.

2.5 The parallel sequencer human machine interface

A real world example of a screen that displays a parallel pump sequence station is shown in figure 2.5.1:

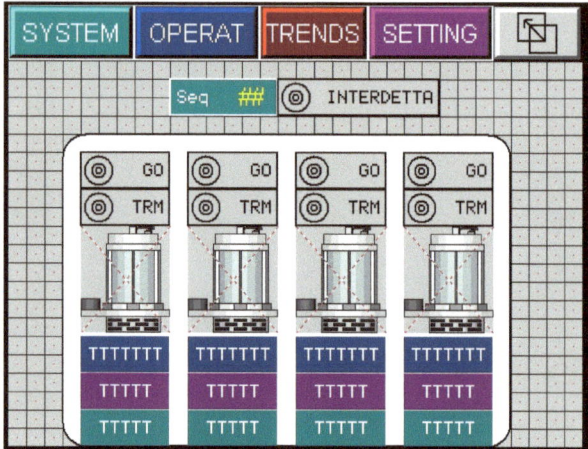

The pump symbol is only visible if it is On; only in this condition the bit 1 of the status variable associated with the pump is active (see block UDFB ElectricMotor).
Other display screens will be detailed in the discussion of the application example.

3.1 Water level measurement and control logics

The example shown is taken from my book "PLC - HMI per Stazioni di sollevamento acque reflue e meteoriche" of the book series AUTOMAZIONE DEGLI IMPIANTI TECNOLOGICI awaiting translation into English.

Logical Development

a) First of all we proceed to the implementation of the measurement of the level in the tank. Let's start by creating a LevelConv object as an instance of the UDFB Conv4_20mA function block, described in chap.3.3. We know that 2 PLC 16-bit registers are required: the first in input to accommodate the register of the reading to be converted; the second output to memorize the engineering value of the level. We will associate the% R201 and% R202 registers of the Plc to the global variables LevelRead and LevelValue. The latter will provide us with the measurement of the level in the tank.

b) We create the function block UDFB Conv4_20mA with the logic instructions shown in chapter 3.3.

c) We define a specific subroutine LevelMeter which, as the first instruction, will call the Conv4_20mA function block through the LevelConv instance, as shown in figure 3.1.1:

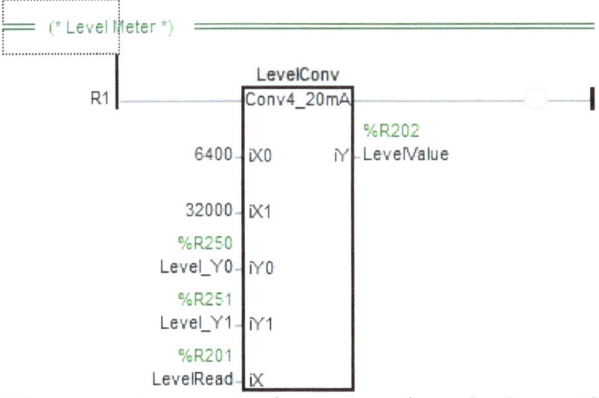

The next rung is required for the final Demo of the application developed in the aforementioned book.

Since we hypothesize to use a level sensor with range 0 - 10.0 m, the values of iY0 and iY1 have been set respectively to 0 and 1000 (10 m with two decimal places) while the values of iX0 and iX1 being native channels are placed at 6400 and 32000 respectively (0 - 32000 or 0 - 32767 if you plan to use different PLC hardware).

c) The subsequent development involves the implementation of control over the acquired level. In this regard, we will create an instance of the UDFB AnalogSts block, already described in chapter 2.4. We know that nine PLC 16-bit registers are needed plus a certain number of Boolean variables. Then we proceed to define the required variables.

The non-retentive global variables used are shown in figure 3.1.2:

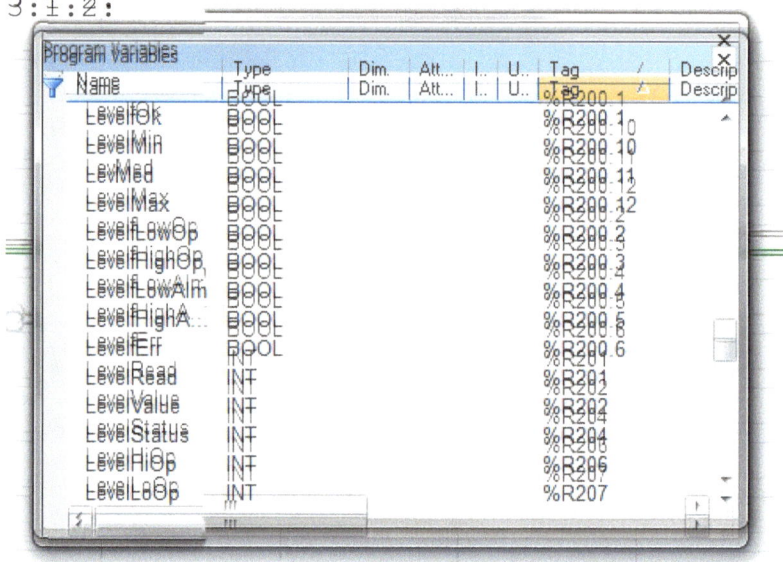

while the retentive ones are 4, as shown in figure 3.1.3:

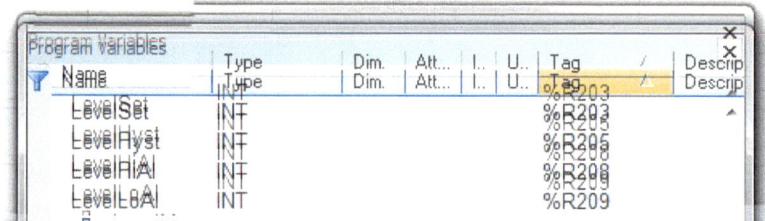

editable by the operator on the HMI keypad.

d) We create the UDFB AnalogSts function block with the logical instructions shown in chapter 2.4.

e) In the LevelMeter subroutine we add the call of the LevelCtrl instance of the AnalogSts function block, as shown in figure 3.1.4:

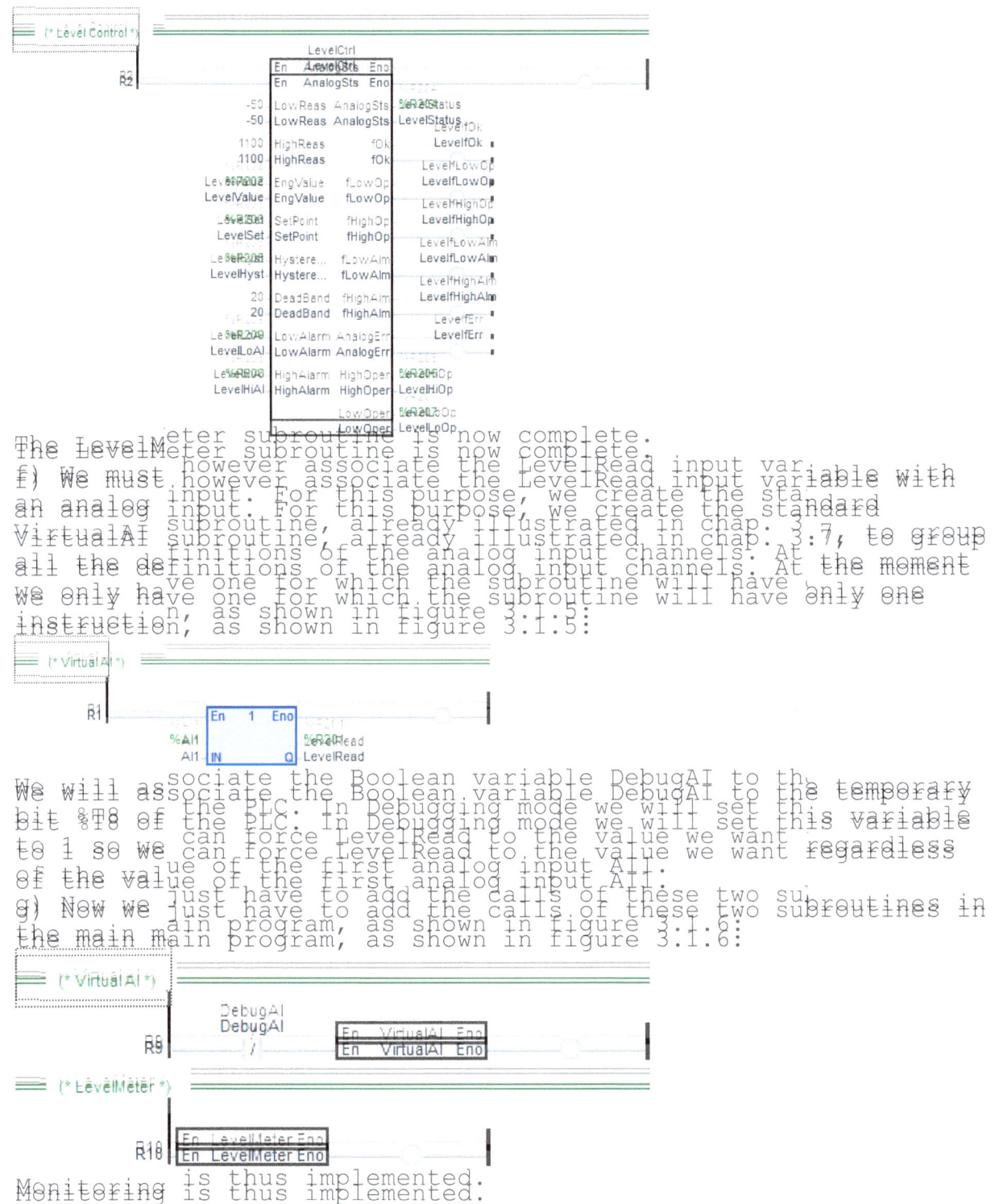

The LevelMeter subroutine is now complete.

f) We must however associate the LevelRead input variable with an analog input. For this purpose, we create the standard VirtualAI subroutine, already illustrated in chap. 3:7, to group all the definitions of the analog input channels. At the moment we only have one for which the subroutine will have only one instruction, as shown in figure 3:1:5:

We will associate the Boolean variable DebugAI to the temporary bit %T8 of the PLC. In Debugging mode we will set this variable to 1 so we can force LevelRead to the value we want regardless of the value of the first analog input AI1.

g) Now we just have to add the calls of these two subroutines in the main main program, as shown in figure 3:1:6:

Monitoring is thus implemented.

3.2 Water level measurement and control HMI

The SYSTEM screen offers a synoptic representation of the pumping system whose automation is described in the previously cited book. The visualization of the level control is shown in figure 3.2.1.

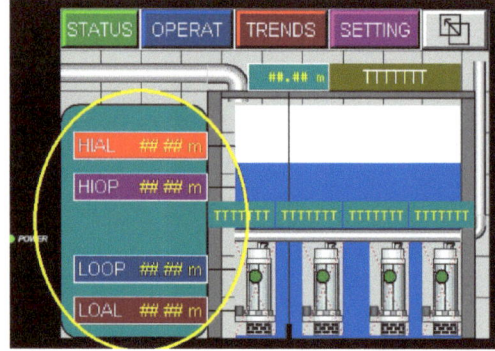

The left side highlighted by the yellow elliptical outline displays, in Numeric Data graphic objects, the values of the operating and level alarm limits. The value of the level and its state are shown in the upper part above the tank, as shown in figure 3.2.2.

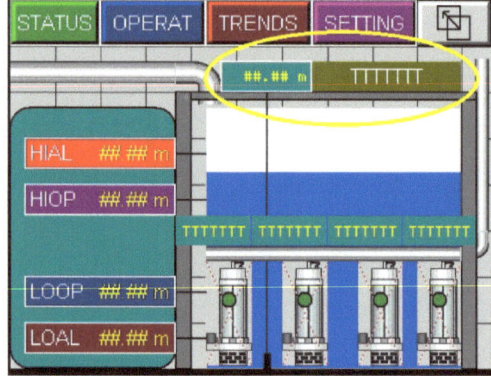

A specific graphic control allows to dynamically visualize the water level, as shown in figure 3.2.3.

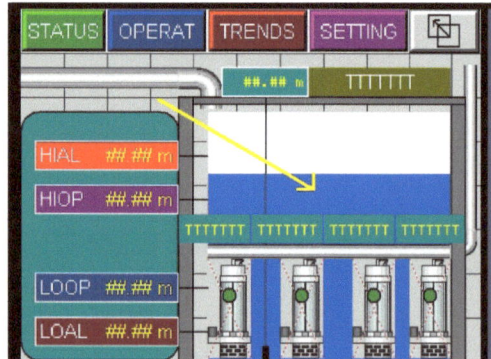

This is a new type of Bar / Meter control that is configured quite simply by indicating both the measured variable LevelValue and the minimum and maximum values that it can reach, 0-1000 in our case, corresponding to the engineering values of 0.00 -10.00 m, as shown in figure 3.2.4.

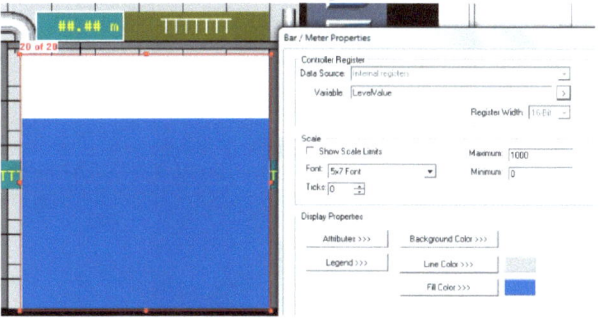

To set the level sensor adjustment parameters dynamically, the SETTING screen is used, as shown in figure 3.2.5:

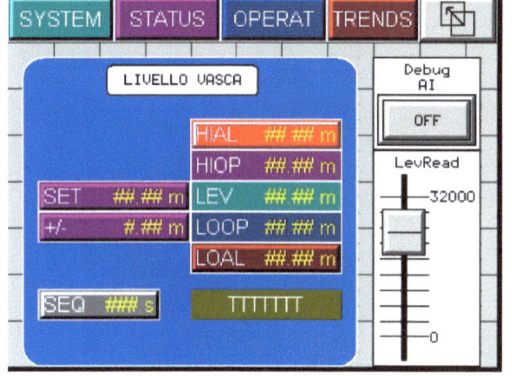

In the blue frame the SET, +/-, HIAL, LOAL values and the duration of the interdiction period are set.
On the right side two tools useful for the test: the AI debug switch and the LevRead slider that allow us to simulate the reading of the analogue input, as shown in figure 3.2.6.

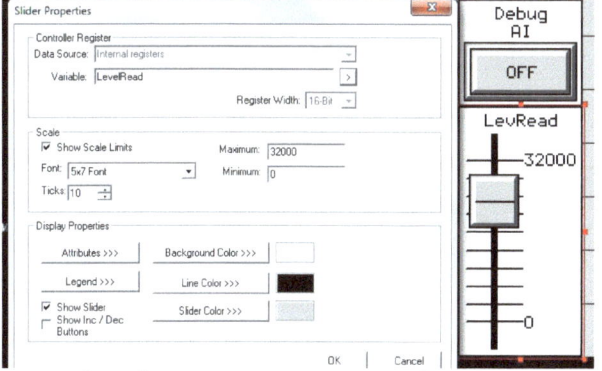

Now let's see some real shots obtained during the test phase in the pumping system automation application described in the book from which the example is taken.
With the slider we start to simulate the actual level to see if the statuses of the level meter are correctly displayed.

We start from the condition of level 0.00 m which corresponds to the status of Low Alarm, as shown in figure 3.2.7:

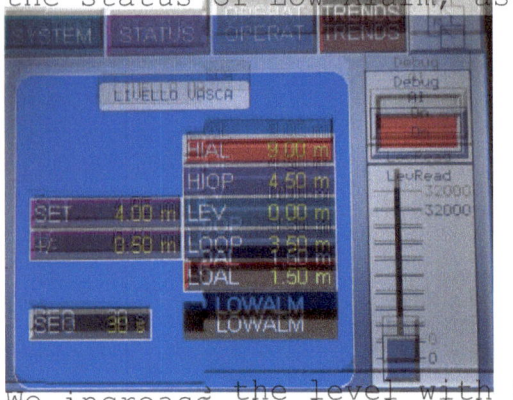

We increase the level with the slider up to 2.44 m corresponding to the Low Operating state, (BASSO in Italian) as shown in figure 3.2.8:

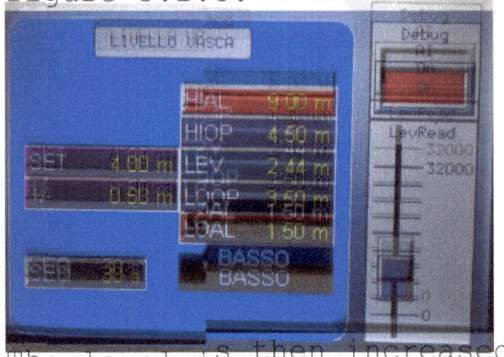

The level is then increased with the slider up to 4.03 m corresponding to the state Ok, as shown in figure 3.2.9:

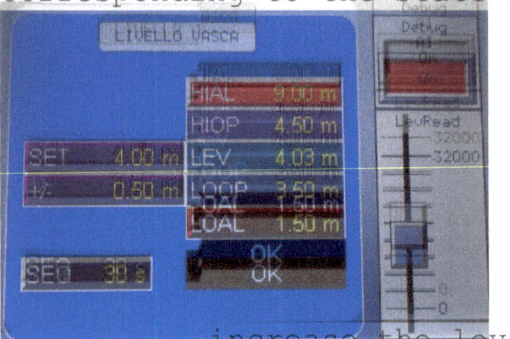

Finally we increase the level with the slider up to bring us to 4.67 m corresponding to the High Operating state corresponding to the HIGH text (ALTO in Italian), as shown in figure 3.2.10:

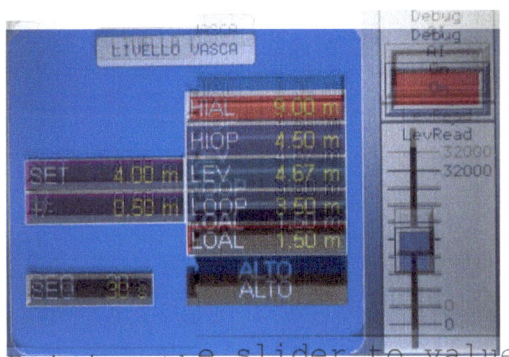

Raising the slider to values close to 32000 will show the status of HIGHALM:

3.3 Pumps sequencer logic

As a next step, we show how the outgoing flags from the AnalogSts function block can be used to implement the functionality related to the sequential start / stop of submersible pumps. If the level exceeds the HIOP value, an available pump will be started. On the other hand, if the level falls below LOOP, one of the active pumps will stop.
The consecutive start / stop will be implemented based on a preset delay "T Station" variable, set on the operator panel, that we will store in the EpSewSeqSet retentive variable associated with the register %R302.
If the level were to exceed the HIAL level, all active pumps would be immediately started and an alarm signal would be generated. If the level falls below the LOAL level all the active pumps would be immediately stopped and an alarm signal would be generated too.
In addition to the control strategy under normal conditions, must be also envisaged the cyclical alternation of the pumps in order to standardize the wear and the automatic substitution of the pump, locked due to failure, with the first pump available.

Parallel Sequencer

To implement the above specifications we will use the design pattern related to the Mot6Seq parallel sequencer that will be illustrated in the third booklet of this series.
a) We insert the call of the Mot6Seq block as a starting line in a specific SewagePumps subroutine, defining the retentive variable EpSewSeqSet, associated to register %R302, as the input parameter setInterdition and the global variable EpSewSeqWord, associated to register% R301, as the output parameter wordSeq, as shown in figure 3.3.1.

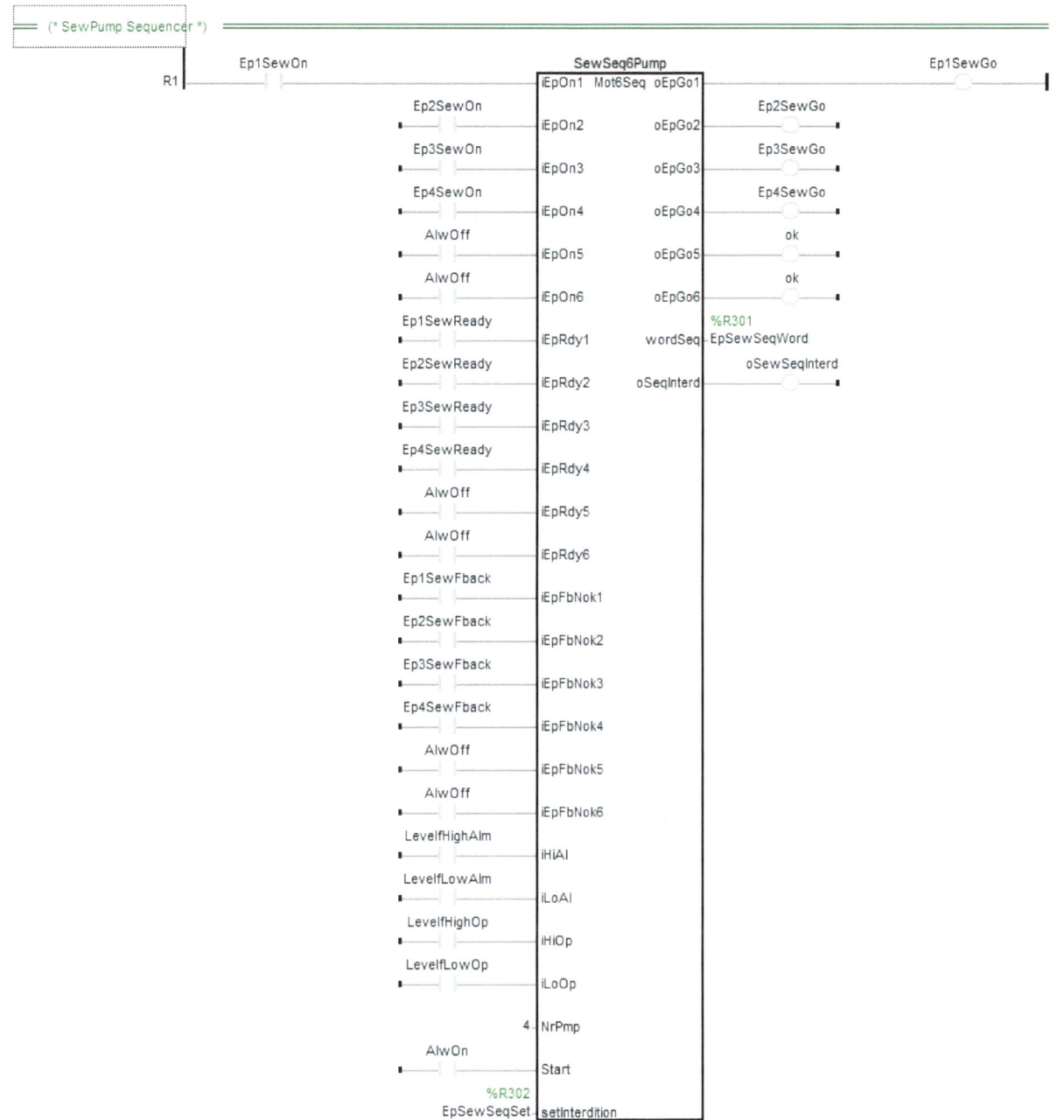

The output flags coming from the AnalogSts function block become the input flags iHiAl, iLoAl, iHiOp, and iLoOP of the sequencer block. The subroutine generates the Go flags for the number of pumps provided in the NrPmp input parameter.
The logic continues normally with the recalls of the ElectricMotor blocks treated exhaustively in the first booklet of this series.
As we can see, the functionality related to the sequential start / stop of the submersible pumps has been obtained using only the functional blocks described in the present series without the need to develop additional user logics.y with the recalls of the

ElectricMotor blocks treated exhaustively in the first booklet of this series.

As we can see, the functionality related to the sequential start / stop of the submersible pumps has been obtained using only the functional blocks described in the present series without the need to develop additional user logics.

3.4 Pumps sequencer HMI

The sequencer display can be obtained with the STATUS screen, which summarizes the operating status, the local commands as well as the Go command variable Go of each pump, also adding the information status from the sequencer, as shown in Figure 3.4.1:

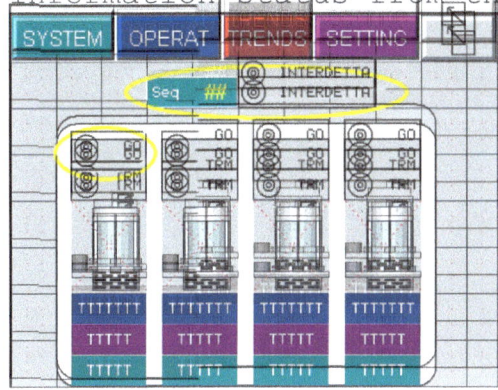

4. Summary

We have reached the end of our third booklet. I would like to thank the reader for the effort made to read it and for the trust given to me as the author.

I am sure that the result achieved will be fully positive and that the quality of the future work carried out, by implementing the techniques acquired, will emerge with sufficient clarity and will be a source of great professional satisfaction.

A sincere wish of good work and a goodbye to the next booklet that will deal with the sequencers.